THE MANY ADVANTAGES OF

THE HOUSE APIARY

Yours sincerely

John Spiller

St. James' Old Vicarage,
    Linden Grove, Taunton.

January, 1952.

THE MANY ADVANTAGES OF
THE HOUSE APIARY

Northern Bee Books

The Advantages of The House Apiary

ISBN 978-1-908904-44-7

There have been extensive unsuccessful attempts to trace the descendants of John Spiller. The publishers are happy to pay an appropriate royalty fee should a descendant make themselves known to them.

Published by Northern Bee Books 2013
Scout Bottom Farm
Mytholmroyd
Hebden Bridge HX7 5JS (UK)

Design and Artwork, D&P Design and Print

Printed by Lightning Source (UK)

THE ADVANTAGES OF
THE HOUSE APIARY.

by

JOHN SPILLER

FOR 32 YEARS MEMBER OF APIS CLUB

21 YEARS DIRECTOR AND 11 YEARS MANAGING DIRECTOR
OF APIS LTD.

FOR SOME YEARS VICE PRESIDENT OF THE BRITISH
BEEKEEPERS ASSOCIATION

VICE PRESIDENT OF SOMERSET BEEKEEPERS
ASSOCIATION

AT ONE TIME SOMERSET DELEGATE TO THE BRITISH
BEEKEEPERS ASSOCIATION

LECTURER ON WILD BEES AND THEIR NESTS
HONEY JUDGE

FOR 6 YEARS PRESIDENT OF THE NATIONAL SOCIETY OF
BEEKEEPERS

MEMBER OF BEE RESEARCH ASSOCIATION

MEMBER OF CENTRAL ASSOCIATION OF BEEKEEPERS

FOR 20 YEARS INTERESTED IN ANALYSING FOREIGN AND
ENGLISH POLLEN AND HONEY

John Spiller, 1880 – 1954

John Spiller in his apiary, 1910

Although born in Bridgwater, Spiller dwelt in Taunton, man and boy. For 35 years he was managing director of Spiller and Webber, Builders' merchants, ironmongers and sanitary engineers of Bridge Street, Taunton. He commenced beekeeping in 1902, running a number of different types of hive, as seen in the 1910 illustration. At about this time he became Hon. Secretary of the Taunton and District Branch of Somerset Beekeepers' Association and became involved in the Association's organisation. In the 1920s he was outspoken in criticism of the parent body, the British Beekeepers' Association and called for disaffiliation. In 1929, when this had been ruled out and Dr Killick had been expelled from the British Beekeepers' Association, he supported him in the establishment of an independent Somerset Society of Beekeepers which was comprised largely of disaffiliated members from Taunton and Williton. For six years he served as its president. Spiller was well known nationally and was a director of the Apis Club, a body of the more erudite and scientifically minded beekeepers. It was said that Webber thought he spent too much time on bees and golf and not enough on the business.

John Spiller

The House Apiary Booklet

Price 12/6 post free.

J. SPILLER
St. James' Old Vicarage
Linden Grove, Taunton

Many advantages compared with outdoor hives.

In semi-darkness of house.

Advertisement from Bee Craft, early 1950s

Spiller began to experiment with keeping his hives in wooden bee houses and he designed several versions which were adaptations of commercially available structures. He encouraged beekeepers to keep their bees in this manner. He was considered to be an authority on this and in 1952 he published a small book entitled "The House Apiary", (IBRA, British Bee Books, a Bibliography, 1500 – 1976, 784, p 227).

Spiller re-joined the Somerset Beekeepers' Association in 1948 and was elected a vice-president, and of the Taunton Division. He died in March 1954, some years after the demise of the Somerset Society of beekeepers.

David Charles, May 2013

THE ADVANTAGES OF
THE HOUSE APIARY.

THIS booklet is written that beekeepers may have the opportunity to know the advantages of the house apiary, its benefits and pleasures for the beekeeper, and the improved conditions for the bees compared with out-door hives,—a wonderful contrast in every way!

To some beekeepers the house apiary may appear extravagant, but this booklet is written to show the merits of the house to those who are not concerned about a little extra cost but desire to know the best and most perfect way for a fascinating hobby.

The cost of the house apiary is not so serious when one considers the amount many keen beekeepers spend on out door hives and also the extra cost of feeding these outdoor hives. The following is a short summary of the advantages.

Owing to the extra warmth in the house, bees store more honey, produce more wax, make better sections of honey and work well on aluminium metal combs.

The house is the only known cure against robber bees while opening hives.

Bees can be made to drift with great advantage for surplus honey.

The Beekeeper can watch his bees at close quarters in the house without being noticed by them.   He becomes more closely associated with his bees and has a better knowledge of the condition of the stocks.

Beekeepers can feed bees with wet wax capping after the honey flow without fear of robber bees.

Queen rearing can be done in comfort! Nucleus hives can be stood at a comfortable height to save the beekeeper stooping without fear of the hives being blown over.

Hives do not require painting, except the alighting boards which are outside the house.

It is very noticeable that empty brood combs in the hives in the house apiary not covered with bees in the winter months keep dry and do not show any mildew growth although this trouble is well known in the outdoor hives.

No hive roofs are required and therefore there is no banging of hive roof to annoy the bees.

Honey can be extracted in house if beekeeper has no other suitable place.

Bees can be fed late in the evening in the autumn when it would be extremely difficult to feed outdoor hives, in the open with wind and rain.

Very much less sugar syrup feeding is required.

Beehives can be attended to at any time convenient to the beekeeper regardless of the weather when it would be impossible to touch outdoor hives, especially in the Autumn when one wishes to give syrup feeding, see sketch page 46 (Feeding out doors).

It is known by observation that the temperature inside the outdoor hives near the walls is only one or two degrees warmer than the air out of doors.

In the house apiary the temperature is 10 to 25 degrees warmer than the air out of doors, so it is obvious the hives in the house must be warmer giving the bees better working conditions and relieving hundreds of foraging bees from staying home to help keep the brood nest warm.

The hives also are less affected by the cold summer nights.   This is proved by the brace comb the bees can make.

Bees are certainly tamer and never vicious.

Bees are less inclined to swarm, probably because of the fairly even temperature in the house.

Evidence of some of the above facts will be given further on and there are more advantages, a few of these will also be mentioned.

For some years I had three house apiaries, one at home in the garden, and two at different places three or four miles in the country. All three houses had different types of beehives, of course, this is not the orthodox way of beekeeping for a beekeeper to have three different kinds of hives, but with the three apiaries far apart, this did not cause any inconvenience and one had a good chance to try the merits of other kinds. In one house I used only the inner casings of my discarded outdoor W.B.C. hives, another house the " National " type and the third house the American " Langstroth " hive manufactured in America. My first house I still have. It was purchased forty years ago after seeing one owned by the late Mr. T. W. Cowan who, as many know was President of the British Beekeepers Association for over fifty years. This house was a portable wooden garage 12ft. by 8ft. with wooden floor. See page opposite. It was converted into a house apiary by fixing a wooden bench 21 inches high along two sides inside the house on which to stand the hives. A height of 21 inches for the bench gives comfort from stooping when opening hives. It is dangerous to have outdoor hives at this height because a gale of wind would blow them over. A useful width for the bench is 2ft. 2ins. giving easy room for moving hives along the shelf without lifting. Underneath the bench there is room for storage of a few empty supers, etc. In the windows of the house bee escapes were made by cutting the glass as explained in drawings further on, so that bees which happen to fly when the beekeeper is opening the hive make for the light of the window above the open hive and get outside the house and return to their entrance again.

Black curtains were fitted to draw over the windows to keep the house dark, but later sliding shutters were fitted which were much warmer and cleaner. The span roof was raised on the side above one of the benches making one sloping roof giving more height for hives to take several supers at one time. The house should be creosoted inside and outside. The creosote inside helps to darken the interior causing all the bees that take to flight to immediately fly to the light outside.

Interior view of No. 1 House Apiary.

The bees' first joy and excitement on apple blossom at the end of April after the long winter.

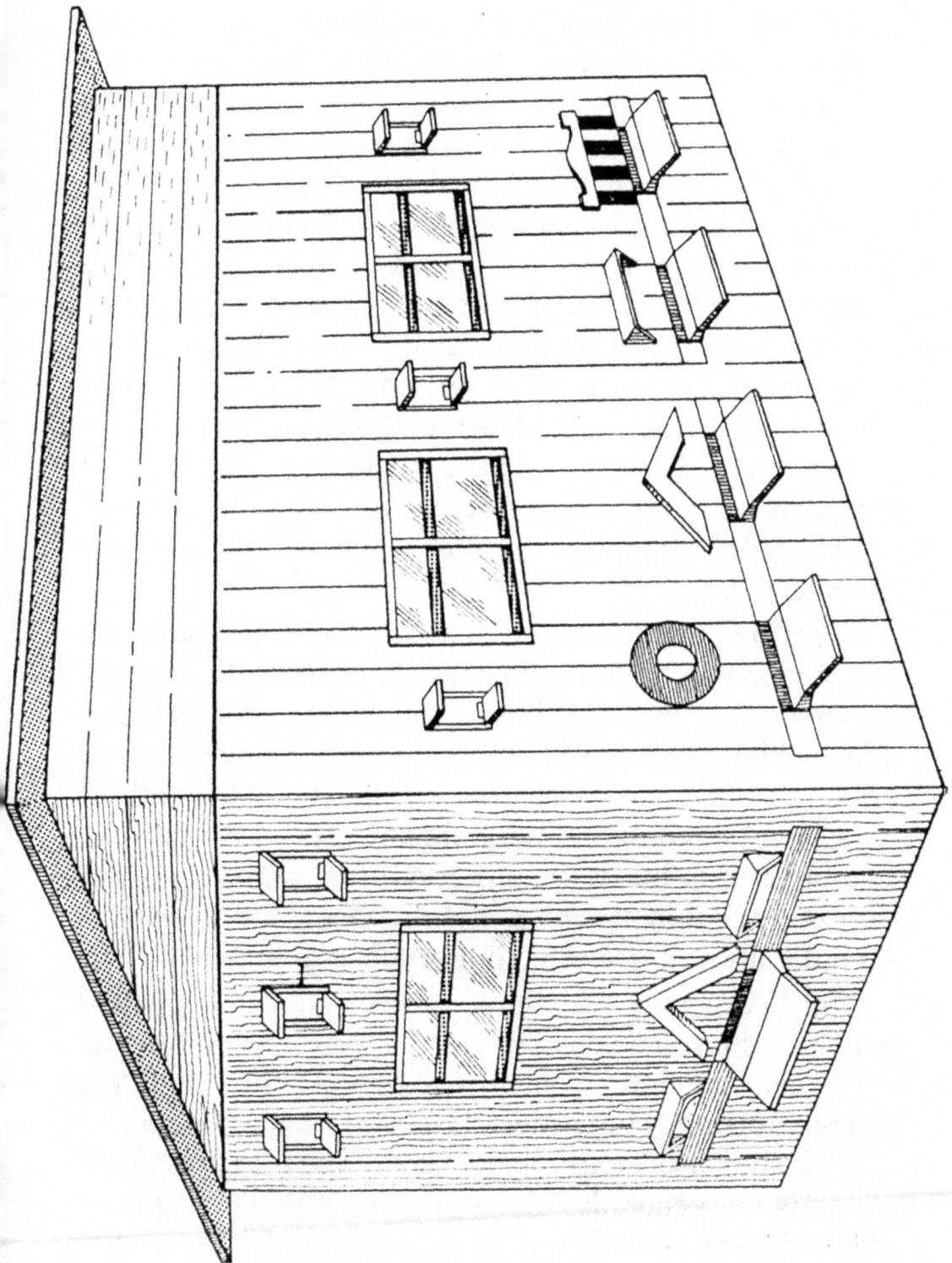

The size house I prefer is 12ft. by 8ft. (No. 1 House apiary) (See drawings). It can take 11 hives, but I only kept two to five hives in this, then there is room for a carpenter's bench, which is very useful in a bee-house, especially a vice for holding a brood frame while queen rearing. The ugly markings over the hive entrances shown on the front of No. 1 house apiary are not considered necessary to-day. These markings were placed on this house 35 years ago when it was imagined that they would help the bees to enter their own home. Other beekeepers with house apiaries have found that this is not necessary. Those beekeepers who prefer markings to guide their bees home should have the markings or colours on the alighting board and not above the entrances.

Experiments have been made to find if bees have any preference for certain colours. It is thought that they prefer deep yellow (such as dandelion) and dark blue. The colour most disliked by them is believed to be pink. Fuller details can be read in other books if the reader wishes to know more about this. It is said bees cannot see red, but it seems certain that they cannot see the exact position of *white* alighting boards, so beekeepers should paint these with colours such as yellow and blue.

The beekeepers who have nice clean white painted outdoor hives arranged to give a good appearance are very keen and interested in their bees, but their bees when returning home most certainly have difficulty in finding their white entrances. A simple way to prove this, which is not generally known, is for those with outdoor hives with the plain fronted type such as the " National " hive to fix horizontally with two drawing pins a strip of black paper one inch above the entrance, the size about 5in. by 1in. They will be surprised to see that the majority of the foraging field bees returning home will pitch on the black paper and walk down to the entrance instead of falling heavily on the alighting board. This shows that the bees can see the black paper easily and prefer to pitch on it rather than on the alighting board.

My second house in the country was 10ft. by 7ft., room for 4 hives on one side and span roof 9ft. high. My third house was 8ft. by 8ft. entirely different from the other two, and there were no windows over hives, only a wooden hinged shutter one end, and the entrance door the other end. The shutter and door left open for light when attending to hives. Also, this third house was raised above ground on brick pillars, so that the hives could stand flat on the floor of house. The disadvantage of this house is that the hives are too low for the beekeeper to stand up when attending them, and he has to kneel on the floor of house. The reason for the hives on floor was to give height above the hives for extra supers, and keep the roof of the house as low as possible.

A friend of mine has a very useful house—11ft. by 11ft. This size is awkward for a garden, but suitable in an orchard. There is plenty of room in the centre to extract honey and in the winter months to store supers, etc. Another beekeeper in West Somerset has an extra large house 30ft. by 12ft. with hives on two sides. He does all his extracting in this and has a carpenter's bench at one end and there is plenty of room to hold beekeepers meetings inside.

A few years ago he invited several beekeepers to see this house and although the weather appeared fairly good when all arrived, a very heavy thunderstorm came

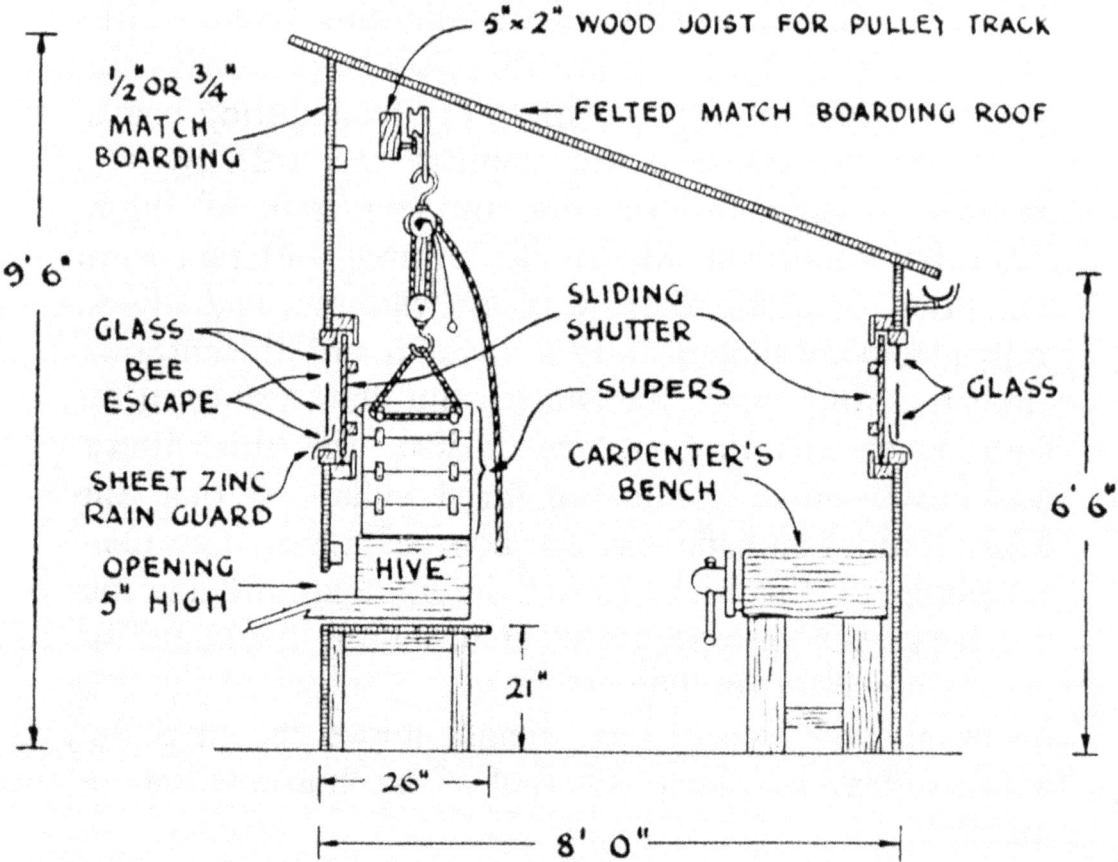

SECTIONAL VIEW OF SERVICEABLE & CONVENIENT HOUSE APIARY

No. 1 House Apiary.

on suddenly, but the meeting of beekeepers went on comfortably in the dry. When the black clouds began to show it was an extraordinary sight to see tens of thousands of field bees returning home. Without the beehouse for shelter this could not have been witnessed.

No. 2 House. Built extra high for skyscraper hives.
one hive with 13 supers 245 lbs. of honey.
„   „   „   11   „   187 lbs. „ „

South end of No. 3 House Apiary showing the wood shutter or wooden skylight which acts as a bee escape while hives inside are being attended to, otherwise the shutter is kept closed. The hives and all entrance boards stand inside the walls of the house and cannot be moved by cattle. The alighting boards shown are either fixed or resting against the house. If knocked will not move or damage the hives.

No. 3 House Apiary.

Showing North end with door, and one hive inside. The outside projecting alighting boards are not fixed to the hives.

When cattle are sent in Orchard to eat the grass, hurdles should be fixed around house as shown in picture, leaving door free for beekeeper to enter.

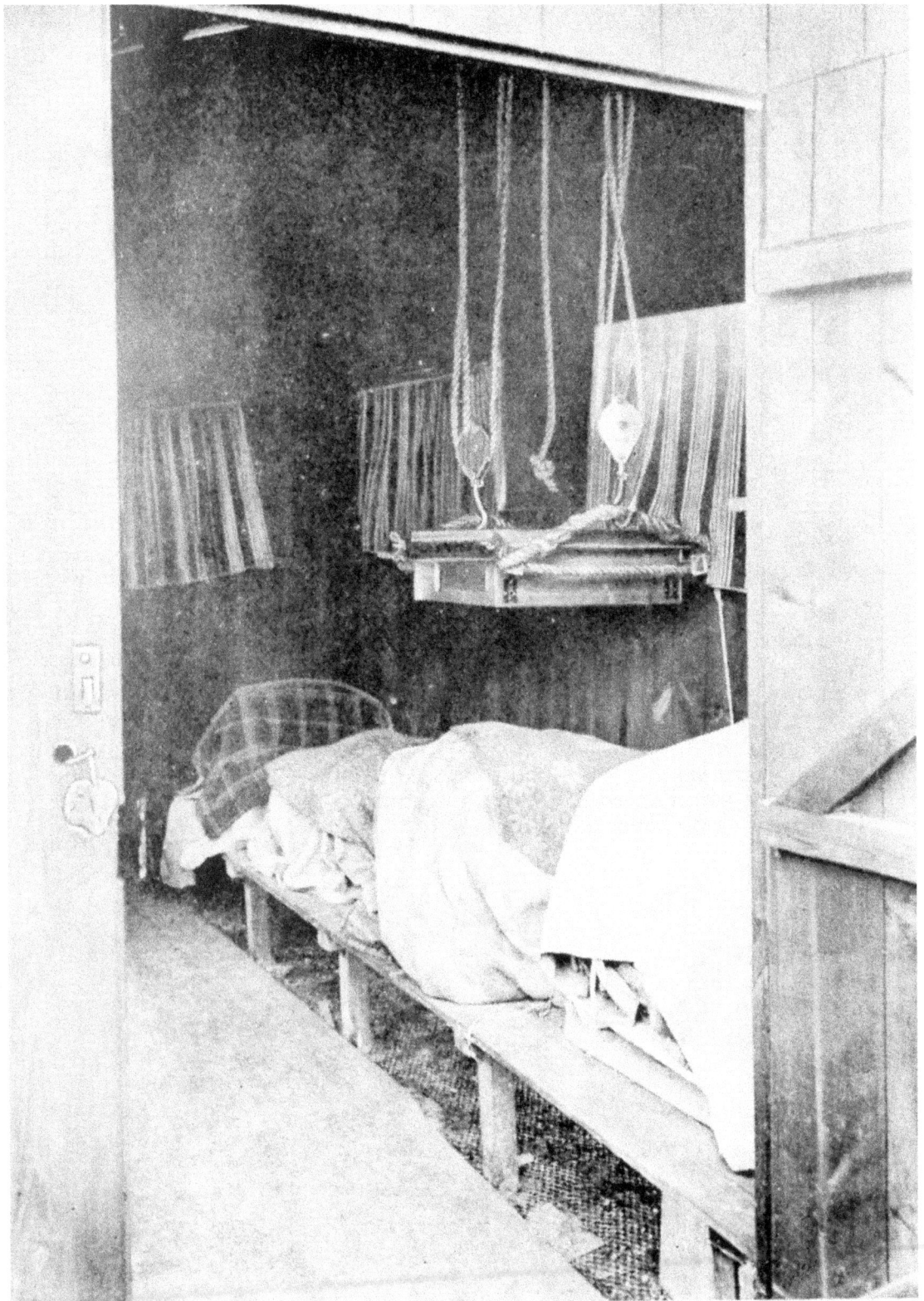

Mid-Winter interior of No. 2 House.

Where hurdles are used there are occasions when the beekeeper wishes to go in front of the hive entrances, this means having to climb over a hurdle. It is more convenient to have a small gate fixed but this must be made extra high as shown here to prevent cattle reaching over the small gate to eat the grass inside. If cattle press heavily against the small gate they will break it down and walk inside and stand in front of the hives and if stung while inside the hurdles the cattle may do damage.

Date 1910.

Another friend made a six sided house with a door on the north side and also another long house 16ft. by 6ft. against the south side of a hedge.

The house apiary must not be confused with a bee hive shed with only two side walls and a front, nor with openings in walls where the ancient straw skeps

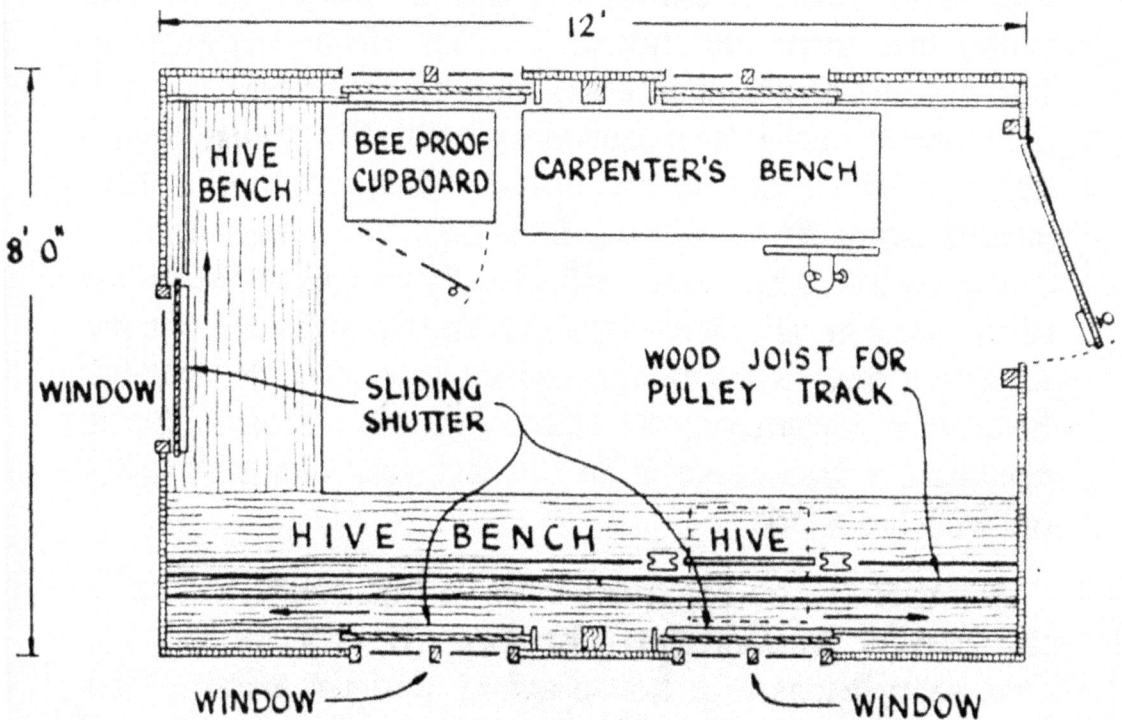

PLAN VIEW OF HOUSE APIARY

No. 1 House Apiary.

stood, nor with the small house shed where there is no room for the beekeeper to stand inside. All these arrangements protect the bees from the weather but give unpleasant working conditions for the beekeeper who is exposed to the weather and neighbouring robber bees.

All beekeepers who have kept bees in house apiaries as well as in outdoor hives say that their best results have been from the hives in the house, and they all consider that the bees in the house are less inclined to swarm. This may be due to the even temperature in the house.

One of the many welcome comforts of the house is that vicious bees become tame and do not attack people who come near the house. They do not appear to know or see the beekeeper while he is opening the hive. It is important to have darkness inside the house apiary and the only light in the house when opening a hive should come from the window over the hive which is being attended to. Bees will not fly in the dark. None of the bees in the other hives in the apiary know or see anything that is happening when one hive is opened. Should it commence to rain while the hive is being opened the beekeeper is in the dry and can still carry on with his work.

Another reason for quiet bees in bee house is due to the careful handling of the bees by the beekeeper. One soon learns in a house apiary that the slightest jar annoys the bees. Even if one only drops a lead pencil one can hear the sudden alarm made in the hives by the bees. Some people, with outdoor hives, are very clumsy when putting on hive roofs and lifts. They do not realise that the noise made inside the hive is tremendous, causing the bees to become vicious. When treated like this bees are on guard at the hive entrance ready to attack any one who comes near. The bees kept in outdoor hives recognise those who open their homes and all human beings are suspected of coming near to annoy them. With the house there is no need to have hive roofs.

During the winter months the house can be used for storing fruit by fixing temporary shelves over the hives. This I have done for years.

Hives and bee appliances kept out in the country miles from home in house apiaries can be protected under lock and key.

I have kept bees for 49 years. For the first seven years I kept them out of doors and therefore know the discomforts of the outdoor hives (See page 44 and photograph 1910).

Several beekeepers have come distances to see my house apiaries and have afterwards written to say it was their red letter day in beekeeping. Many have copied the idea and made house apiaries of their own.

It is surprising to me that the house apiary is so little known in the British Isles as it is so very suitable for our uncertain weather. In other countries house apiaries are used. In central Europe some beekeepers have too many stocks in one crowded house with two rows of hives one row above the other on two or three sides in the house. This means that there are too many bees in one place. Each district has its limit of nectar supply for bees. The beekeeper having a row of hives above those on the ground shows that he does not expect his hives to want many supers to each hive. Houses filled like this make very torturous work for the bee-keeper as there is very little room for him to move. It is good protection for the bees from the weather but certainly not a pleasant hobby for the beekeeper. A large number of stocks of bees in one place means all these extra brood nests using up very much of the available nectar in the district. Each brood nest requires a large amount of food for the young.

One expense and trouble with outdoor apiaries is having to keep the grass and weeds cut around the hives. This work annoys the bees. Some beekeepers go to the expense of a concrete or tarmac floor to stand the hives on and this saves the beekeeper from standing on damp grass while opening a hive.

There are many people who cannot have a house apiary for various reasons, probably there are local restrictions against fixing up a shed in one's garden or elsewhere. Permission has to be received from local authorities to put up even small sheds in case there is to be an increase in the rates for such a shed, but house apiaries come under the rules of Agricultural or Bee farming and are probably not rateable.

One may already have a garden tool shed. Perhaps this could have one part converted for beehouse, keeping this part of the house dark inside.

One friend of mine used a hay loft over a disused stable. Another beekeeper kept his bees in the roof of his house. He found the bees preferred the extra summer heat under the roof. One of his hives supplied him with 100lbs. of surplus honey. In the warmth and fairly even temperature night and day, the bees are not hindered from making the wax comb freely. If one's home is in the middle of a terrace of houses, where the appearance is the same all along the terrace, the bees are very liable to lose their way home if there is nothing conspicuous to guide them. It is necessary to paint or whitewash one chimney near by. Those who own a car and can run to and fro 3 or 4 miles into the country will find that farmers are usually very pleased to allow the use of an odd corner on their farm for bees, but go to a good farmer as good farming is good for beekeeping.

Amateur beekeepers in Britain keep bees as an enjoyable and interesting hobby and the extra moderate expense of running into the country is not considered.

When the house is kept in the garden, (this also applies to outdoor hives), it is much better to have a 6ft. high wood fence, wall or hedge in front of the hives, 6ft. or 7ft. away from the bee entrances, so that when the flying foraging bees go out it keeps them flying above this height and away from people in the garden, and the bees do not see people in garden from their hive entrances.

Some think that a house apiary is a permanent fixture. This is not necessarily so. My three house apiaries were moved a total number of eleven times. These houses had wooden floors. The one I have now at home stands on a concrete floor, so of course this could only be moved to a new place without the floor: but a concrete floor is better than one of wood.

Some who know very little about the house apiary are prejudiced against it and say " bees rob "! but the house apiary is the only cure known to prevent robbing. In countries where it is the custom each year to build up stocks to high " skyscrapers " it is known that the house apiary is the best preventative against robbing while opening hives, to add new supers. With outdoor " sky-scraper " hives there is the great risk of attracting robber bees.

Hives used in a house apiary can be made of cheap material and without hive roofs and need not be painted as they are not exposed to the weather. Only the inner casing of W.B.C. brood chambers and supers need be used for cheapness, but the National pattern or the American Langstroth hive is easier to work and the expense is not much more. The photograph opposite shows the inside of my No. 2 house in mid-winter. These hives, as may be noticed, are made up of the inner cases of the W.B.C. hive and the floor board. They are shown covered up with wraps for extra warmth. Note the empty super suspended from the rope pulleys, and the running track for lifting several supers all at once when adding a new empty super in the nectar season. Sad to say since I have not been able to buy the good strain of bees from abroad, I have not had occasion since 1940 to use the pulleys. They were used every year before this. The house apiary is much more suitable for the use of pulley blocks and running track for lifting bodily in one movement several supers screwed together with iron plates (See page 15 for fixing iron plates), than for the outdoor hives. Some beekeepers use this method out doors but while the supers are suspended in mid-air, they are exposed to the outside cold air and especially at the end of the honey flow are attacked by robber bees. This does not happen to the suspended supers in the house apiary. The chief advantage of the pulleys is the saving of time when adding a new super which can be done in a few seconds without disturbing the bees. It is done so quickly,

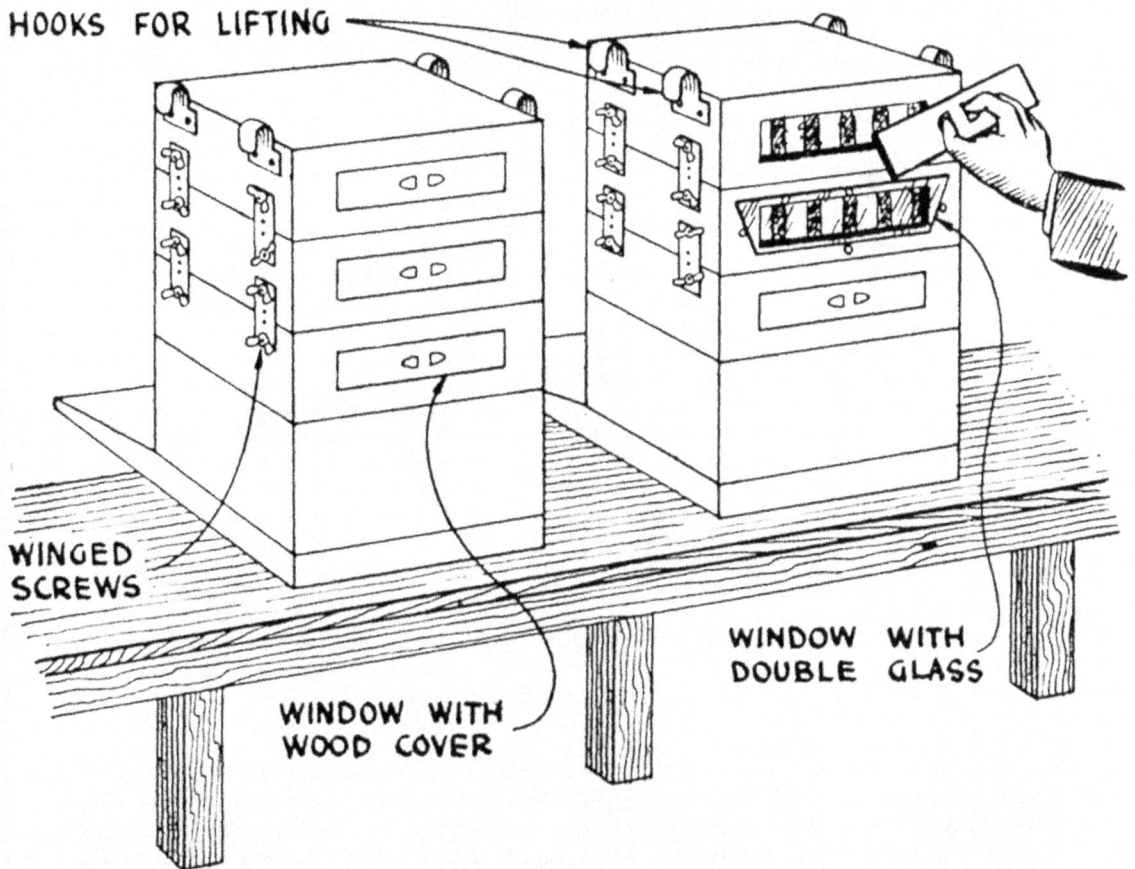

HOOKS FOR LIFTING

WINGED
SCREWS

WINDOW WITH
WOOD COVER

WINDOW WITH
DOUBLE GLASS

almost without the bees knowing anything has happened, the supers are kept warm and should there be any brace comb between the frames, so often found between the upper and lower frames in the supers, it saves the bees from repairing the breakages, whereas when having to lift off the top supers one by one it causes confusion amongst the bees and may unsettle the bees for some hours before they begin work again. To work the pulleys with comfort it is better to have two pulleys, one at each side of the hive, otherwise when only one is used the heavy suspended supers may spin around. Rope must be used for the pulleys because iron chains cause a noise and may knock against the hive alarming the bees. The most suitable pulley blocks to use are as shown in

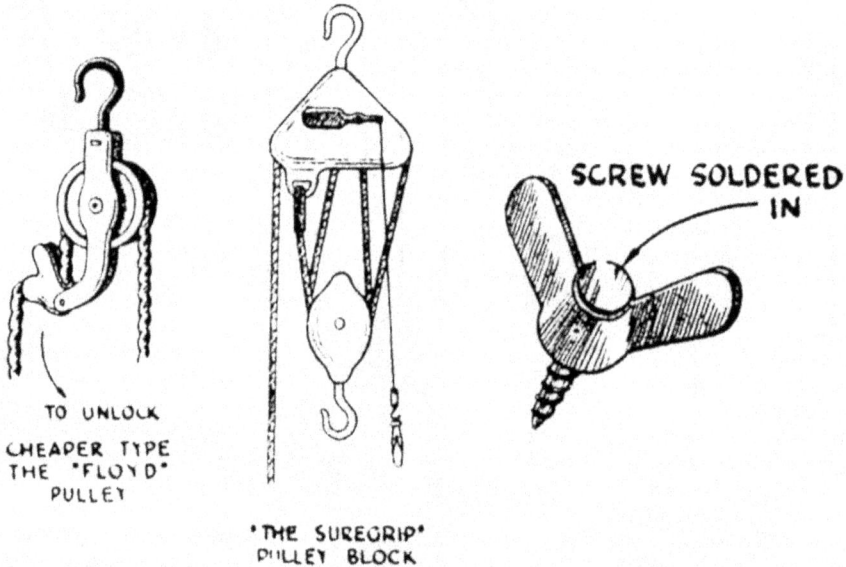

TO UNLOCK
CHEAPER TYPE
THE "FLOYD"
PULLEY

"THE SUREGRIP"
PULLEY BLOCK

SCREW SOLDERED
IN

As can be seen in the drawing when using pulleys the top supers which are to be lifted are screwed together with iron plates size about 3″ long by ½″ or larger. These are usually bought with four screw holes each but it is only necessary to use two screw holes, the top and bottom hole for fixing the iron plates. If the hives stand fairly close to each other it becomes awkward to fix the screws in with a screw-driver. It is much easier and quicker to use screws fixed with solder in winged nuts (see drawing).

the drawings. The " Suregrip " and the " Floyd ". The " Suregrip " pattern I have used for 35 years. The " Floyd " is a cheaper type.

There is on the market in America a portable lifting machine on wheels for lifting supers. This is suitable both for the hives in the house apiary or outside.

This photograph taken in mid-winter makes it difficult to imagine the activity that will take place in a few month's time. Two years previous to when this photograph was taken at the beginning of the summer there were only two stocks of bees similar in size to those shown, and these two stocks gathered four cwts. of surplus honey, an average of two cwts. per hive. There were thirteen supers above the brood chamber on one hive and eleven on the other. The whole summer there were no signs of robber bees while the new supers were being added. My friends were able to stand inside the house and look through the windows at the back of the supers and see for themselves the sealed honey comb. By standing in the semi-darkness of the house they could see the foraging field bees returning to their hives 30 or 40 a second to each hive. In fact, every day it was almost impossible to count them. If these same hives had been out of doors, the visitors could not have gone near these bees because of robber bees attracted by the opening of the hives when adding new supers. One wonders how much less honey the hives would have produced if they had not been protected by the house, and also whether they would have swarmed.

Even if a house is large enough to take a dozen hives of bees, it is better for the amateur to only have four or five hives. The amateur will enjoy his beekeeping far more and his bees can be attended to in peace. In the forty years I have used house apiaries I usually have had only four or five stocks of bees in each house and for many years up to 1941 usually averaged over 80lbs. of surplus honey per stock. I feel sure if I had kept double the number of stocks in each house apiary I should have had very much less surplus honey but more work and the trouble of taking off half empty supers. One must remember there are other neighbouring beekeepers in the district if near a town and their bees also collect nectar. During the last war after 1940 several people in my district took up beekeeping hoping to get some honey to compensate the shortage of sugar. The result of this was that my bees only gave about 35lbs. of surplus honey per hive instead of 80lbs. and many of the beginners did not get any. I must say in fairness that part of my loss was caused by the poor quality of queen bees available in this country during the war and several clover grass fields ploughed up for growing corn. We could not buy from Italy because we were at war with that country and the Government would not allow us to send money to America to buy good American-Italian queens.

Queen Rearing.

The bee-house is especially suitable for the fascinating study of queen-rearing, for the work can all be done inside the house at times when it would be quite impossible out of doors. The temperature in the house varies according to the kind of weather at the time, but it can easily be 10 deg. to 25 deg. Fah. higher inside than outside. It is considered safe to graft queen grubs if the temperature is over 70 deg. Fah. and if the bee-keeper prefers a higher temperature he can warm the house with an oil heating stove. If the beekeeper is opening a nucleus hive with a virgin queen or has a queen

from another district which has never been out flying
in the district he must be careful to close all bee-escapes
in the windows and keep the door of the house closed
and locked on the inside to prevent any one opening
door, otherwise the queen may make for the light and
get lost.

Experts say all British beekeepers should rear their
own queens, but they do not warn them of the cost and
of the many disappointments for the novice. It is an
interesting and pleasant experience to try to rear a few
queens if one can afford the cost and time. This booklet
is to introduce the value and advantages of the house
apiary and does not deal with beekeeping for the
beginner nor with queen rearing. There are many books
obtainable on these subjects, but the most wonderful
book on queen rearing that I have read was published in
1946 by Mr. L. E. Snelgrove. The author has gone to
much trouble to warn beekeepers about the many
causes of failure in artificial queen rearing. If a book like
this had been published many years back, we older
beekeepers would not have heard our friends saying one
to another " How I nearly reared a queen " ! One of
the many warnings mentioned is that of the dragonfly
which is overlooked and unknown to many beekeepers
but I presume nothing can be done about this danger.

When I tried to rear queens, I was very pleased to see
the beautiful dragonflies hovering, darting forward and
backward in my garden, not knowing at the time they
were only there after my virgin queens. In my garden I
have over thirty fully grown trees so it is easy to see the
dragonflies against the background of the foliage of the

trees. An American beekeeper in one year considered he lost over sixty virgin queens by dragonflies so to continue his queen rearing had to move away from the district.

Queen rearing in the house can be commenced early in April if the weather is good but of course there is the risk of the weather changing before the queens mate. English queen breeders are liable to have bad weather even in Mid-Summer. In 1944 the weather in West Somerset was good early in April and the young queens were laying when the hives were examined on May 1st. It was an exceptionally early year but very few years are like this.

There is a good case for the house apiary where a stock of bees has swarmed and the beekeeper thought he had cut out all the queen cells except one, but has over-looked a second one. His mistake will soon be known in a bee house when the two unhatched virgin queens are piping at each other from their cells. He can then act promptly in the way he wishes.

THE WEIGHING MACHINE.

A hive kept on a suitable weighing machine in a house apiary is a great source of interest. The hive can be weighed in comfort in any weather and the weight is not affected by the rain. My machine is a platform type with steelyard. This was given to me by a friend, a much appreciated present. Before the war these were sold for invalids to weigh themselves each day in their homes. A variation of weight can be measured as little as 1 oz. (See sketch on page 21). Those of the spring balance type are very unpleasant and troublesome to use. It would be practically impossible for a beekeeper to use this pattern several times a day whereas the platform type takes only a second or two each time to weigh the hive, and does not jar the hive and worry the bees inside.

Sketch of weighing scales with bee hive.
Most suitable to use in bee house.

During a good honey flow it is interesting to see the hive increasing in weight every hour. Sometimes in the morning a hive will increase 1lb. per hour. In the middle of the day, the hive loses weight owing to thousands of young worker bees and drones out flying (4,800 bees to the 1lb. weight). As evening approaches the hive regains its weight, some days it will increase by 9 lbs. I have known a hive increase by 7 lbs. in weight by eleven o'clock in the morning, but after this came a thunder-storm and stopped all work. The hives lose weight during the night owing to the bees fanning the moisture away from the unripe honey. I have known them lose as much as $2\frac{1}{2}$ lbs. in one night when lime nectar was coming in.

By weighing the hive every day, some years one can notice the hive has not increased in weight for a week or two in the middle of the summer. One wonders what is happening, whether the source of nectar has stopped or whether the bees are preparing to swarm, or if super-sedure of the old queen has commenced, and the bees are not inclined to work until the young virgin queen is mated and laying. A strong hive during the winter months usually loses about 1 lb. in weight each month until the middle of January, but when the queeen starts laying the losses begin to increase. From October 1st to the end of February a loss of 6 lbs. only for the five months. Outdoor hives unless they have been given very special protection would lose considerably more weight than this. I gave the following details in the " Bee World " of 1942 Summer Graph, April 1943 and it may be of interest here.

DAILY WEIGHT OF HIVE OF BEES.

The chart gives the weight of a medium size stock of bees each day from May 1st, to August 11th, 1942.

The vertical lines represent each day and the horizontal lines give each 2 lbs. in weight.

Up to June 25th, owing to poor weather, it seemed as though there would not be a honey harvest. However, at this date the weather really set in fine and at the very same time the lime trees began to bloom. During the next 18 days the hive increased in weight 49½ lbs. For five of these days (marked " K " on the chart) when the lime trees were giving off nectar at their best, the weight went up 32½ lbs. gross, but on each of the five nights there was a loss of 2 lbs. to 2½ lbs., making the increase in five days 22 lbs. net. After this, clover nectar began to come in and the loss at night stopped. By this it appears that the lime nectar is very thin and watery as compared to clover nectar. The amount of clover nectar collected shows up rather poor compared with the lime, but this was owing to the clover being half a mile away from home; however, in the next 21 days, on six days of which the bees could not work, the hive increased 18 lbs. By August 4th the hive had increased (from May 1st) to 90 lbs. net weight. From June 25th to August 4th, 67 lbs. net. The total surplus honey taken off was 70 lbs. of extracted honey, and also there was plenty of winter stores left on, although the bees were given 4 lb. of syrup for a little extra.

It may surprise many beekeepers who spend most of their summer cutting out queen cells that from April until the middle of August the hive was not opened, except for placing on supers and taking them off. The brood nest was not seen once. The queen was in her third summer and the stock did not swarm.   This would have shown by the weight of the hive dropping 5 lbs. or 6 lbs. in one day; also it is certain without referring to the weight. The hive was kept in a bee house and all the supers were fitted with glass windows showing at a glance the thousands of bees inside. When a swarm goes off, very few bees are seen near these windows for the next two or three weeks.

It was found in August that the old queen had been superseded. Five other hives were left in the same way as the above with old queens and treated the same by not touching the brood nest for the summer; none of them swarmed, but all superseded their queens. These six hives had to be left with old queens in 1941 as it was impossible to buy sufficient queens from queen breeders. The fact that none of these stocks swarmed, even with old queens, must certainly prove the importance of buying queens from firms who breed from selection and with the best scientific method. It takes a war for one to risk keeping old queens for their third summer and to find out additional good qualities, and also to realise the blessing it is to be able to buy queens from abroad in peace-time, where the weather conditions are ideal for queen breeding. The English breeders have great difficulty in supplying the demand, owing to the shortness of the summer and the uncertainty of the climate. The surplus honey from the above six hives averaged 64 lbs. per hive, but other hives, where young queens were introduced, the surplus was 86 lbs. per hive.   This shows the advantage of re-queening every year with well-bred queens. The cost is small compared with the extra surplus honey and the liability of swarming is much reduced.

Details of the lettering given on the Chart :—

A.   Good weather, apple blossom out; 10 lbs. increase in 7 days.

B.   May 6th.   First super put on, some new sealed honey seen along the top of brood nest.

C.   Cold and rainy.

D.   Only one good day.   Turnip field in flower grown for seed, one mile away.   Bees could be seen returning home covered with turnip pollen.

E.   May blossom out.

F.   Cold and showery.

G.   Hot weather, 92 deg. in the shade; raspberry nectar; 14 lbs. increase in 6 days.

H.   Cooler again; lost 8 lbs. in 8 days.

I.   Good weather but very little nectar for bees. Probably this was the period of queen supersedure which was found at the end of the Summer.

J.   Very good weather; lime tree in flower.   The first 8 days increased 27 lbs. net; in 18 days, with partly clover nectar, increased 49½ lbs. net.

K.   Five days increased 32½ lbs. gross, but 22 lbs. net.

L.   Clover nectar only, but half a mile away; 14 lbs. increase in 12 days.

M.   August 4th.   Highest point reached, 90 lbs. net.

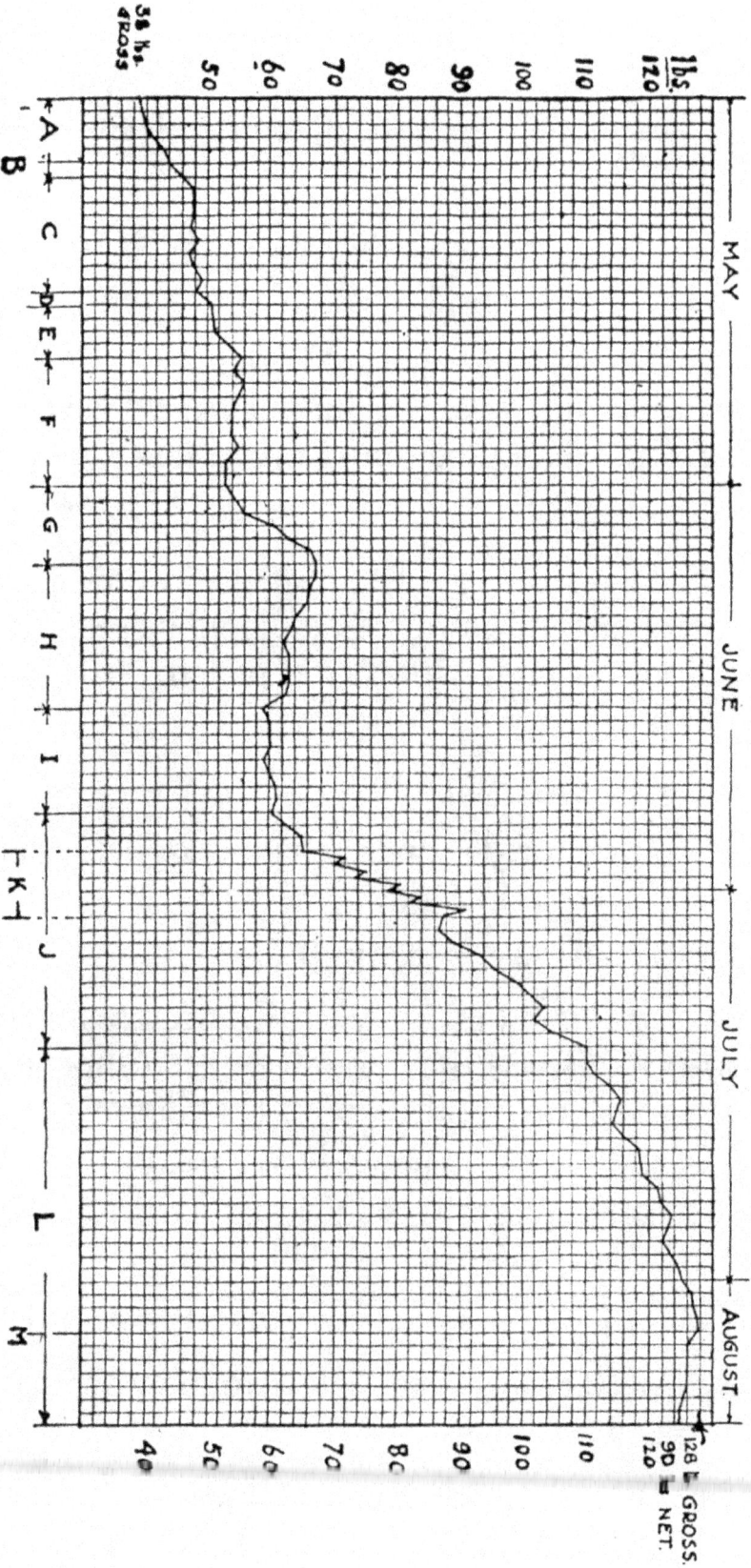

HONEY FLOW GRAPH 1942

The hives in the bee-house can be opened for spring-cleaning without risk of chilling the brood three or four weeks earlier than the out door hives when it would be extremely foolish to open outdoor hives. At the end of the summer honey flow opening outdoor hives attracts robber bees. Opening hives in a bee-house, does not attract the robbers.

When spring-cleaning the house apiary one should see that all spider webs are brushed away from inside and also from the outside of the house. Spiders can soon build their webs inside and outside in the many odd corners and during the summer trap some of the bees and even virgin queens.

The entrances of the hives are best if facing south and west. The east is not so good owing to the bees flying in early Spring when the weather is good and those bees returning home in the afternoon when the sun is not shining on their entrances. If they do not happen to alight straight into the entrance of the hive but fall to the ground with their heavy load of nectar and pollen, they very soon become too chilled to rise and to fly up to their entrances. If the hive entrance is on the north side the casualties are greater. These chilled bees if picked up one by one and placed in a small box and warmed in one's pocket or elsewhere, will soon revive and fly to their homes. Hundreds of bees can be saved in early spring by doing this. One extra worker bee in the hive in early Spring is worth more than a dozen bees in a few weeks time.

When the hives are placed 6in. to 7in. from wall of the

bee house as mentioned further on, the house receives the scent of the nectar that the bees are collecting. This is very interesting and as the summer goes on, so the scent of one source of nectar changes to other sources. It is good to have the scent of the hives like this, should one hive become infected with foul brood, the bee-keeper would smell it at once and deal with it, before it spread to other hives. When the bees are collecting nectar from the gooseberry the bee-keeper is likely to be alarmed by a suspicious smell of foul brood, but in a few days this smell disappears when the bees stop gathering from the gooseberry. With outdoor hives it is impossible to notice foul brood until the disease is in a rotten state, unless the bee-keeper happens to open the hive in time.

One soon finds when opening hives in a house that it is not really essential to use a smoker, nor to wear a veil, but perhaps it is safer for some to wear a veil. For many years I never used a *smoker*, only requiring a carbolic cloth kept in proper order to drive bees from off the tops of the frames, before putting on the top quilt. Of course this may be that for well over thirty years, my bees were imported pure Italians which were always in the same gentle mood. In those days I usually averaged 80 lbs. per hive or more.

The warmth of the bee house may have helped this, but one reason was the quality of the pure bred queens. I am afraid these days with the kind of bees we are now allowed to have, bred in this country, in uncertain breeding weather, we are very lucky if we get half this amount of honey.

VISITORS IN HOUSE APIARY.

The sketch on the next page shows how one can open a stock of bees in the house and look for the queen to show visitors. There is no need for bee-veils and the young lady with bare arms is pointing to the queen.

It is very foolish to open a hive like this in mid-Summer, as it slows up the work of the bees and inclines them to prepare for swarming, but when a bee-keeper's meeting is held or on a special occasion one of the hives may be opened to show the visitors.

What a contrast to a meeting held in the open when bee-veils have to be worn by everyone and need careful adjustment in case the wind blows the loose netting of veil against the face. Recently a lady at one of these outdoor meetings was stung on the knee when the wind blew her skirt and she was so frantic she tore off her under garment to get rid of the bees.

I have had quite 20 people at one time packed in my bee-house, while demonstrating with an open hive, and very few bees leave the combs to fly, and these few bees make straight for the daylight and pass out through the escape in the window over the hive. None of the people were anxious to wear veils.

In a house it is a saving of time to have all supers fitted with a glass window at the back, either with one pane of glass and a wood cover, or with double glass windows to prevent loss of heat. The bees' normal activities can then be seen at a glance, without disturbing the bees in any way and no loss of heat from the hive caused by opening the hive and disturbing the bees.

At a glance the beekeeper knows when a new super is needed, and also whether the bees have swarmed and

View of visitors in bee house without any fear and not anxious for bee veil. See page opposite.

gone. It may be mentioned here for interest that it is considered, that bees work better in the supers when

Another discomfort for beekeepers and for the bees in outdoor hives.

When opening a hive in warm Summer weather the beekeeper has to use a beekeeper's smoker. As all beekeepers know to their annoyance it often goes out the moment he wishes to use it and to avoid this he tries vigorous efforts with the bellows to keep it alight. This causes smuts to form in the bellows and these smuts are blown into the hive on to the lovely white honey comb and later the bellows get so hot that fire is expelled from the spout of the smoker and singes the wings of some of the bees in the hive.

But worst of all for the beekeeper he perspires so much under the bee veil that immediately after the work he has to have a hot bath and clean dry underclothing.

there is a window with double glass in the super above the brood nest.

The best covering for the top of hive is the well-made American enamelled cloth which the bees cannot nibble. This has been considered the best covering for quite 80 years, but was not obtainable during the war years. It is an extra joy to have one or two hives covered with Burgess' glass quilt, but this should always have an extra pane of glass resting on the top to prevent condensation under the bottom glass of quilt. Visitors can then be shown the bees at work on the tops of the frames, without affecting the bees.

If a glass quilt is on a hive that is to be opened and if brace comb can be seen through the glass do not prise up the glass quilt, otherwise the glass will break, but turn it slowly left to right to and fro to break the brace comb from the glass. It is a wonderful experience for visitors to see the bees at work under the glass quilt, making the brace comb and filling the cells with honey. This pleasure can only be seen in the house apiaries. Of course, the bees need some extra warm covering over the American cloth or glass quilt.

THERMOMETER UNDER GLASS QUILT.

Another advantage with the house apiary is that a thermometer can be used under the double glass quilt during the winter, when it would be difficult to read one in an outdoor hive. The clock faced thermometer must be used as these work automatically and do not require shaking. The ordinary mercury thermometer may not register correctly as it must be laid horizontally under the glass quilt. The clock faced thermometer is laid face upwards in the feed hole of the bottom glass quilt immediately next to the cluster of bees. The vital part of the thermometer next to the bees must be protected from the bees with perforated zinc so that they cannot spoil the working of the thermometer by smothering the cold metal parts or glass with wax and propolis, as explained further on.

If a stock of bees is in perfect condition the temperature will keep at 59 deg. during the winter months, but the temperature of the hive rises in very cold weather as the bees must eat up extra stores to make more warmth. It just shows how much more the outdoor stocks of bees will consume being more exposed.

Watching the thermometer in the house does not affect the bees at all, but with an outdoor hive removing the hive roof andwarm quilts would disturb the bees.

THE POSITION OF HIVE ENTRANCES.

The hive fronts can be allowed to stand flat against the inside wall of the house No. 1, but in this position the bees cannot be watched from inside the house, nor does the scent of the nectar enter the house. The best position is as shown in drawings No. 2 and 3 with space of 6in. to 7in. between the inside wall and front of hive. This allows plenty of room for the scent of the flowers from which the bees are gathering nectar to enter the house. Also in the winter months the hive entrances cannot be blocked up with snow and the woodpeckers and tits are afraid to venture inside to eat the bees. See drawing of " Tunnel " page 42. The beekeeper with this space can watch at close quarters all activities at the hive entrance from inside the house. He can see the foraging field bees returning home with the various colours of pollen, and the guardsmen and scavenger bees, and the undertaker bees carrying out the dead. Also very interesting to the bee-keeper to watch the outgoings and returning of the virgin queen without the virgin queen knowing that he is near by.

We all know that in warm weather some of the young bees stand outside at the entrance beyond the group of " fanning bees " and stay in the hot current of air that is being forced out of the hive. With the outside hive one cannot be close enough to watch carefully without

being attacked by the scout bees, but with the house apiary the beekeeper can look down on these young bees and if preferred use a hand magnifying glass, and see one bee grooming another bee and others attacking a wasp at the entrance, or paying attention to a " dancing bee" which appears to have pain in its abdomen, rubbing its hind legs against its lower parts, and the beekeeper may wonder if this trouble is anything to do with the working of the wax glands.

In May, 1944, I was fortunate enough to catch one of these dancing bees at the hive entrance, and found a wax scale in one of its pockets. This confirms my article about dancing bees in the Dec. 1943 " *Bee World*."

This dancing seen in the house apiary is entirely different from the wonderful accounts of dancing bees given by Professor von Frish and Ph. D. J. Schiller. In a house apiary the beekeeper is able to see much that passes unnoticed by owners of outdoor hives.

In May, 1944 I caught one of my dancing bees carrying pollen on her hind legs, she arrived home and was dancing at the hive entrance. When the pollen was examined under a microscope, it was Trifolium incarnation clover. This crimson clover is grown in the early summer for feeding cattle, and as soon as the flower comes, the farmer cuts it, so the bees do not get many days to gather its abundant yielding nectar, which is of great value to the bees in early summer, therefore one can understand why the bees who find a field of this will return home and dance.

With a more general use of the house apiary, perhaps more will be known about the secret ways of the bees.

These dancing bees seen at the hive entrance can also be watched through the glass quilt on the top bars of the hive under the glass. Sometimes one solitary bee standing alone can be seen dancing and wagging its abdomen. These bees cannot possibly be trying to attract the attention of other bees. The bees at the hive entrance do not appear to be worrying about gathering nectar. It may be they are resting after foraging. It is possible that they are young bees at the age when their wax glands begin to secrete wax naturally. The wax scales while congealing in the wax pockets probably adhere slightly inside the pockets causing the bee to use her hind legs to rub hard and try to dislodge the wax scales. It is known that sometimes bees have difficulty in dislodging these scales. These wax scales are so transparent that they cannot be seen in the ordinary way. This is mentioned to show that the beekeeper with a house apiary has a better understanding of the bees than those with outdoor hives.

There is nothing in beekeeping to equal the delightful experience of the sound of the singing roar of the hives in the house apiary. This is a great pleasure to the amateur beekeeper. On a warm summer's evening in the house, when there is a good honey flow the wonderful joyous tune from the large number of fanning bees at the entrances and the lovely scent of the nectar coming in from all the hives reminds one of the harvest festival hymn which is sung about the waving golden cornfields. " That even they are singing." Here we can understand the beekeeper's imagination, the singing tune of the roar of the hives caused by the bees fanning their wings and giving a unique song of the harvest of the nectar safely stored away.

When there is much fanning at the entrance in hot weather and the hive has several supers on, the hive can be raised off the floor-board half an inch to give better ventilation. In a bee-house the bees are not affected by cold nights, nor disturbed by robber bees. When the door of the house is opened by anyone entering the house the bees may fly out to the light from the raised hives, flying sometimes up against the person entering, but this can be prevented by placing perforated zinc over the $\frac{1}{2}$ inch opening on the three sides. When the bees do not require the extra ventilation they fill up the perforated holes with propolis! When hives are raised like this the front entrances have extra wide and high openings, exposing the brood nests directly to the outside air should the weather become cooler, and thousands of foraging field bees would then have to stay home to keep the brood warm. Small blocks of wood of suitable size should be placed at each end of the entrance to narrow the openings.

Beekeepers with outdoor hives often have the entrances too wide in the Summer, and robber bees quietly crawl in to help themselves to the honey, and the beekeepers rarely get any surplus when this happens. In the Autumn the robber bees are very different, and then they become vicious and anxious for winter stores.

With reference to the openings in the house for the hive entrances. Some beekeepers with a house apiary just fit their hive entrances in the most simple way by cutting a hole through the wall of the house wide enough to allow the front projecting floor board through the wall as drawing No. 1. This may suit some beekeepers but does not allow freedom to move hives

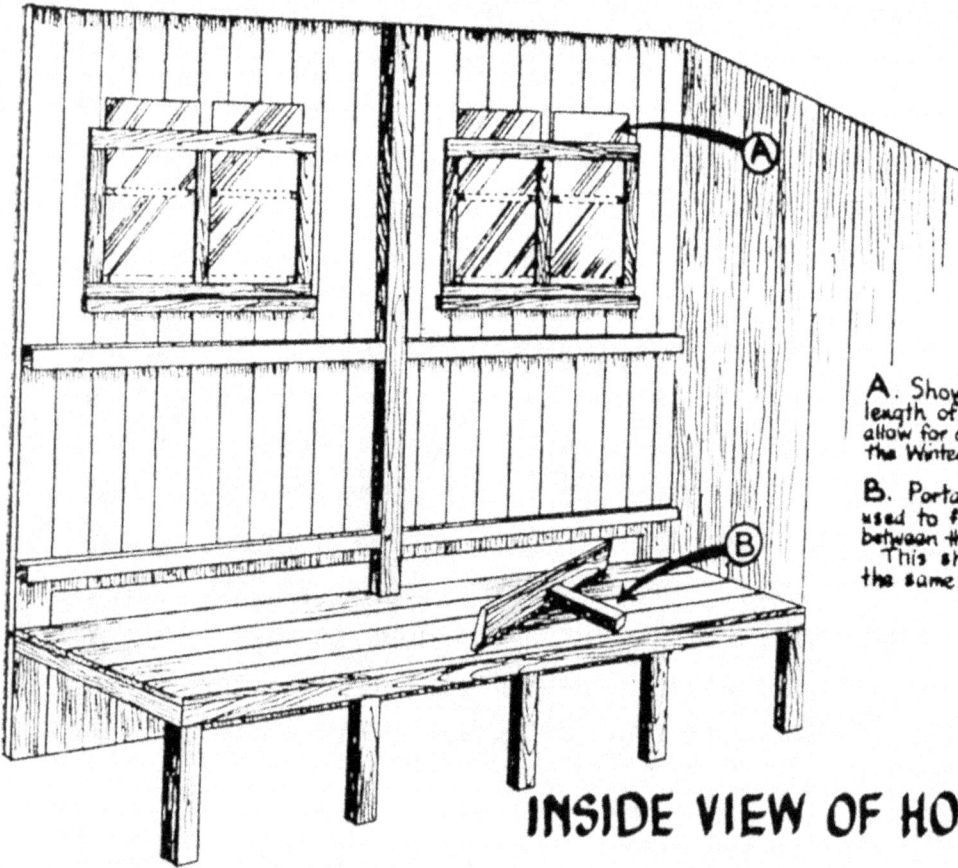

A. Showing the extra length of glass, to allow for closing up in the Winter

B. Portable shutter used to fill the openings between the hives. This shutter should be the same width as hive.

INSIDE VIEW OF HOUSE

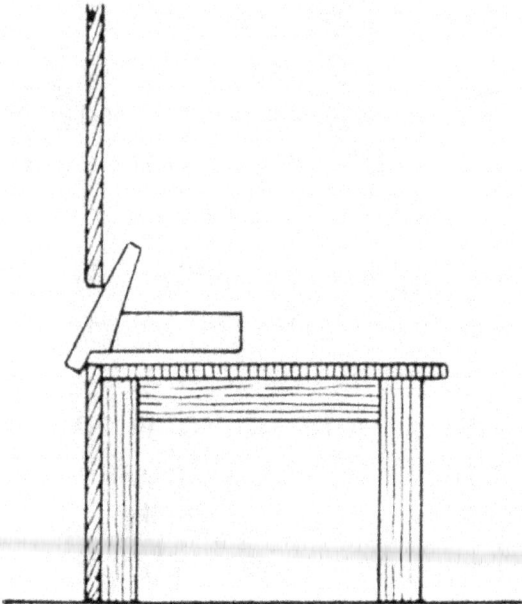

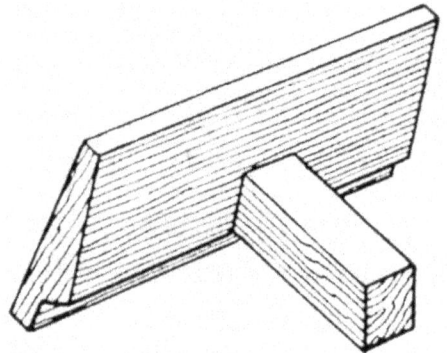

PORTABLE SHUTTER. A quick and simple method for closing wall opening for hive entrance and spaces between hives to keep out light.

slightly sideways when necessary and also in Summer
supers are too close to the wall of house. I prefer to have

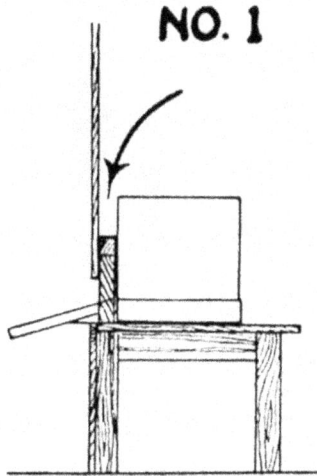

NO. 1

No. 1 Hive entrance not recommended.

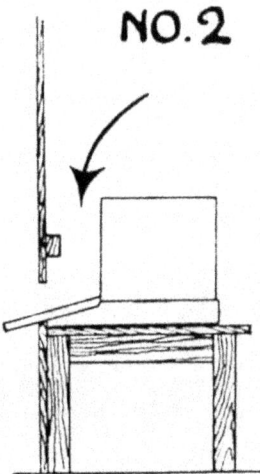

NO. 2

No. 2 Good position.

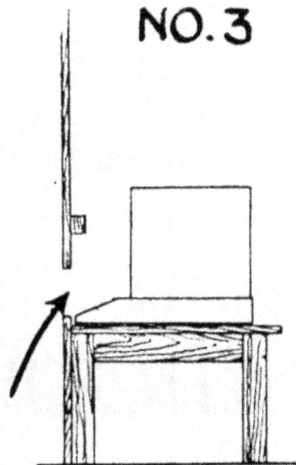

NO. 3

No. 3 Similar to No. 2 except that the
alighting board is inside the house out of
touch by cattle. If an extra extending
alighting board required should be fixed
to the outside of wall of house as shown in
photograph of No. 3 House.

one long opening the whole length of the side of the
house 5 in. high except for the necessary wall standard

supports every six feet. The 5 in. high opening if the
floor boards are not more than 2½in. high. (See drawings).

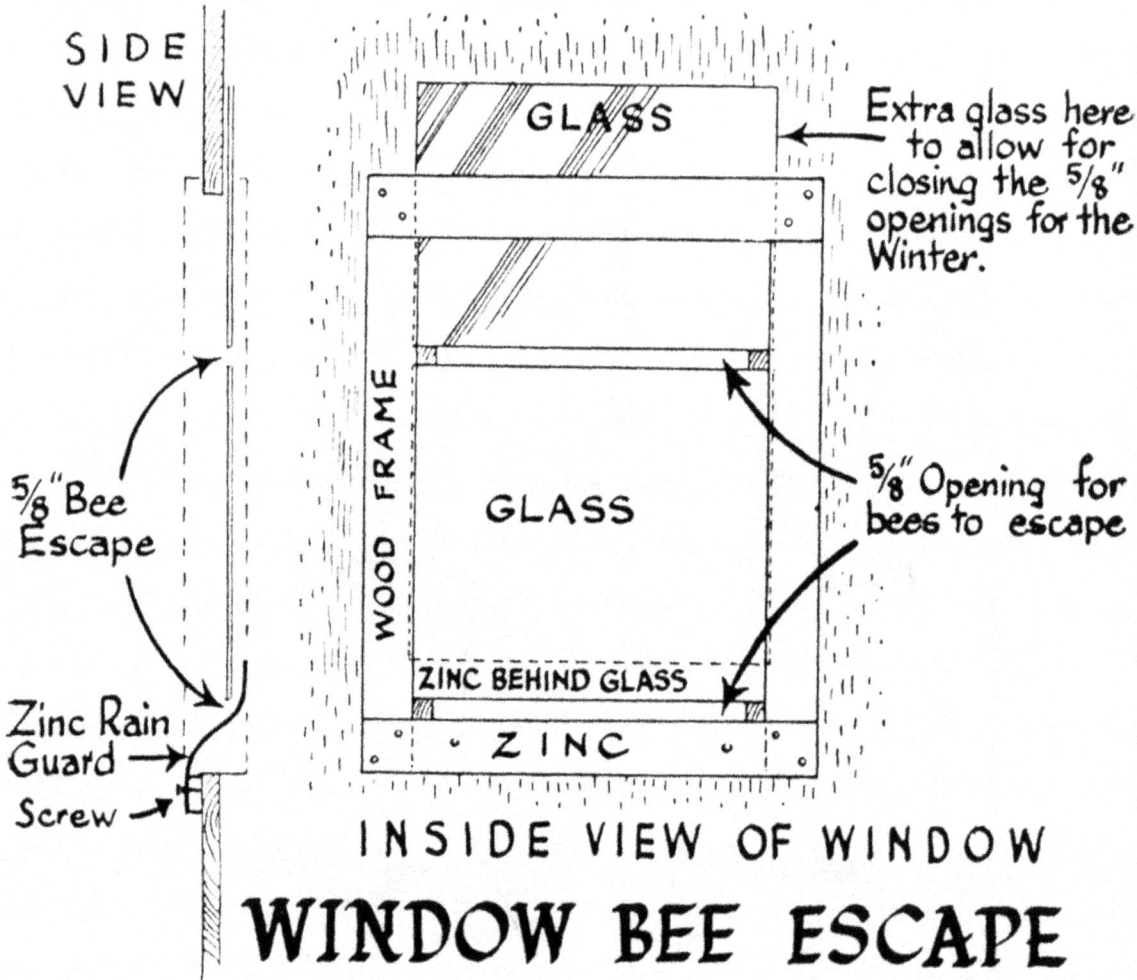

SIDE VIEW

GLASS

Extra glass here to allow for closing the ⅝" openings for the Winter.

WOOD FRAME

GLASS

⅝" Opening for bees to escape

⅝" Bee Escape

ZINC BEHIND GLASS

Zinc Rain Guard →

ZINC

Screw →

INSIDE VIEW OF WINDOW

WINDOW BEE ESCAPE

   With reference to the bee-escapes (see drawing), the
simplest way is to have in the glass window two ⅝in.
openings the width of the glass, one at the bottom and
the other two-thirds higher up the window and sufficient
glass at the top to allow for sliding the glass down to
close the two openings during the winter months, to
keep the house warm where curtains are used instead of

shutters and also can be dropped when dealing with young queens. If the house is exposed to wind and rain there is an improved rainproof bee escape introduced

RAIN PROOF BEE ESCAPE

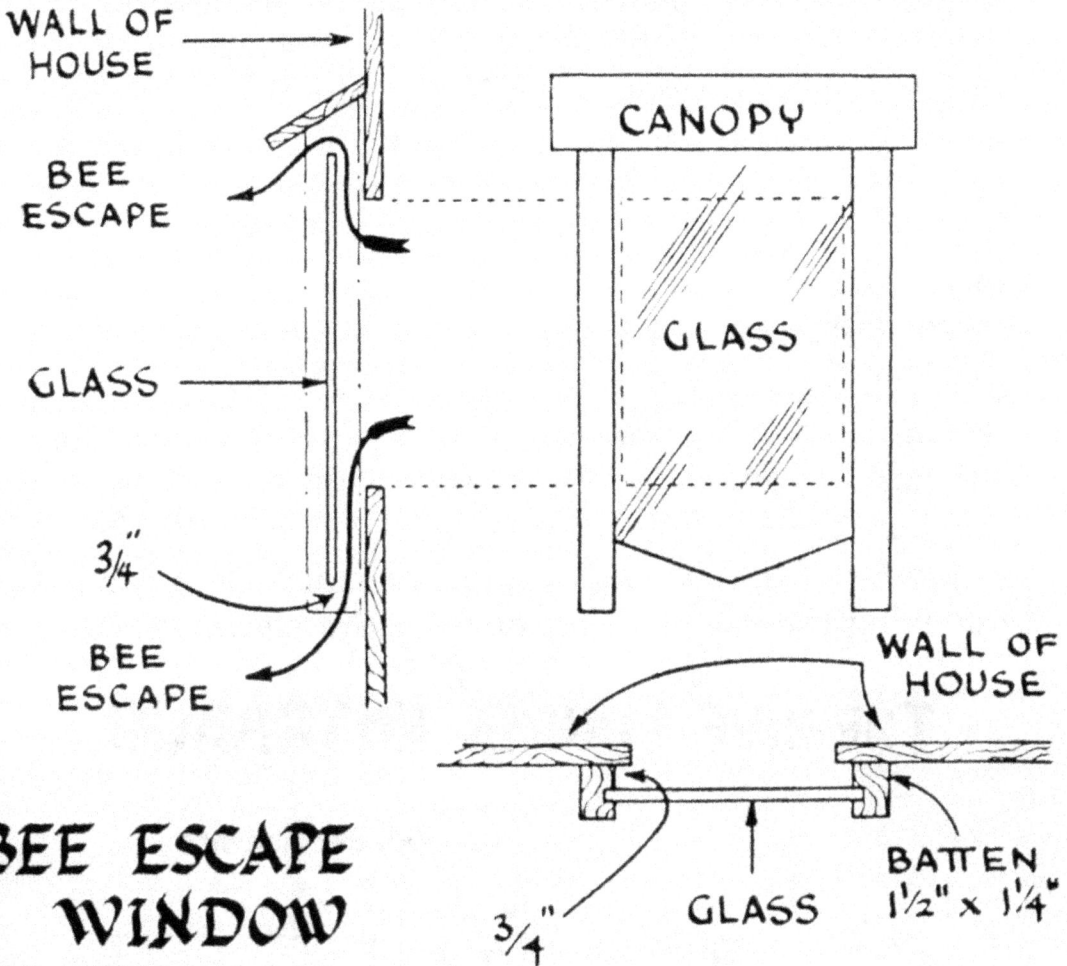

WALL OF HOUSE

BEE ESCAPE

GLASS

CANOPY

GLASS

GLASS

3/4"

BEE ESCAPE

WALL OF HOUSE

BEE ESCAPE WINDOW

AS INTRODUCED BY MR. L. J. HARTNELL OF TAUNTON

BATTEN 1½" x 1¼"

3/4"

GLASS

by Mr. L. J. Hartnell of Taunton which can be seen in one of the drawings.

During the late autumn and winter months we have

to guard against field mice entering our stocks of bees, especially when in the country. Most beekeepers know what to do about this with their ordinary out door hives such as placing a flat piece of zinc the whole length of the hive entrance with a slot cut out of the centre bottom edge 7ins. long by $\frac{3}{16}$in. high., the zinc fixed with drawing pins against the front of the hive. Mice cannot pass through $\frac{3}{8}$in. space, but the field mice are pleased to enter inside the house apiary. Precautions must be taken to keep them out or trap them by setting one or two " break-back " spring mouse-traps in the house during the winter.

When the mice get in the house apiary for the Winter, they eat holes and nest in the warm covering wraps on the hives. I have used two devices to keep them out, although I have always been successful with these by keeping the mice out of the house, I feel it is simpler to use mouse traps because of the work and time needed for fixing them. Sketches of these are shown here. I

THE HIVE "TUNNEL" ENTRANCE

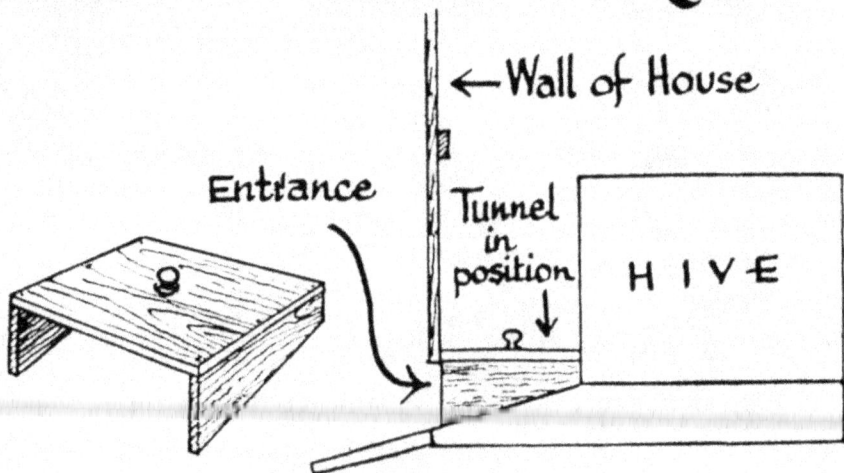

←Wall of House

Entrance

Tunnel in position ↓

HIVE

have named them " Blinkers " and " Tunnel ". It is
hardly necessary to mention that the mouse traps used in
the country house apiaries should be the " Break-back "
trap or any kind that will kill the mice at once, because
the beekeeper in the Winter months may not visit the
houses for many days. The " Tunnel " pattern is
dropped in between the front of hive and the inside of
the wall of the house and stands on the alighting board
of the hive and then the hive is pressed tightly against
the wall of the house. This " Tunnel " pattern is a
good protection against the birds, such as tits and woodpeckers. The " Tunnel " stands too low to allow the
woodpeckers to get in near the hive entrance where any
disturbed bees in mid Winter may venture to come out
when hearing the tappings of the woodpecker and the
" Tunnel " is too dark for the tits to venture in far

HIVE "BLINKERS"

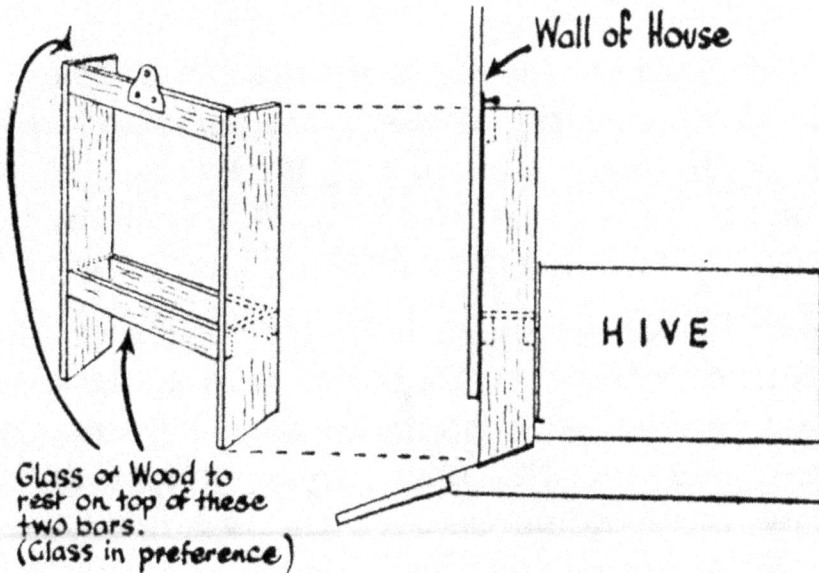

Wall of House

HIVE

Glass or Wood to
rest on top of these
two bars.
(Glass in preference)

enough.   There is danger of snow closing up the " Tunnel " entrance so the beekeeper must be on the watch to remove any snow to prevent the bees being suffocated.   With the " Tunnel " pattern it is not so useful for the beekeeper because he cannot see the bees from inside the house as can be done with the " Blinker " pattern.

One Winter, not using any of these devices in one of the houses in the country I trapped over 20 field mice. I usually set 3 traps on the floor.   It is very rare that mice enter my home house apiary.   The trouble is in the country.

I heard of a sad case from one of my friends who took a hive to the heather on the Quantocks.   On arriving found he had not enough hive lifts to cover up the supers so he placed a waterproof wrapping on the top of the hive to cover the supers and tied up everything, but evidently left a place where mice found a way in, on the top of the warm supers.   When removing the supers after the heather was finished, out ran a mouse and inside there was the mouse's Winter nest with 29 acorns.   The beekeeper was very grieved to disturb the nest and it was a great mystery to him how the mouse (or mice) had carried the acorns up to the top of hive.

This photograph shows one of my outdoor apiaries as it was over forty years ago.   A great contrast to my beekeeping now with house apiaries.   It may be of interest to say that this photograph was specially taken to show me standing between the two first stocks of bees that gave me over 100 lbs. each of surplus honey,

Date 1910.

the seventh year of my bee-keeping. Since having house apiaries I have had several dozen stocks of bees producing over 100 lbs. of honey in one year. My four best " sky-scraper " stocks gave as follows : 245 lbs., 187 lbs., 160 lbs. and 146 lbs. I attribute this success to the house apiary and also to the pure Italian bees.

I used to re-queen each year, end of the Summer, except where the queen happened to be extra good. While re-queening one can appreciate the value of the house when looking for the old queen. There is no fear of robber bees nor of chilling the brood during this operation.

It was surprising to me that I did not get any surplus honey the first three summers of my beekeeping and I began to think that I was unfortunate and living in a barren county unsuitable for bees. Now, I wonder how many cwts. extra honey my bees could have produced if I had kept them in a house apiary in my first 9 years instead of outdoors exposed to the cold nights, wind and weather causing the bees to use up honey to keep the brood nest warm. Also as bees require more winter stores to carry them safely through the winter than those kept in a house apiary it is impossible to estimate how many cwts. of sugar have been saved by having the bees in the house apiary.

By way of an experiment a nucleus hive of bees with only three brood frames of honey and brood have lived safely through the winter in a house apiary owing to the warmth inside.

Another discomfort for the beekeeper with outdoor hives—Some stocks need winter stores. This has to be done in the Autumn when the weather may be wet and windy.

BEES DRIFTING.

It is well known that bees kept in outdoor hives standing in a row will drift to the windward end of the row. This is no detriment if the bees are free from disease. After keeping bees in house apiaries for nearly forty years I cannot say I have noticed drifting in this way.

At the most I have only had four hives in a row in a house. Bees returning home can see the house and know if they belong to one end or the other or the centre of the building.

BEES CAN BE FORCED TO DRIFT.

With great advantage when the house is placed in a special position. I have done this for several years with one of my house apiaries in the country. (See sketch). The house was built on the east side of a small thickly wooded copse to protect it from the west winter winds. I also fixed for extra protection a 6ft. high wattle fence about 7ft. away from the west side of the house. The north was closed in between the house and the wattle fence to save the house from the strong winter rain and wind. Only four stocks of bees were kept in this house all with their entrances facing west. At the beginning of the spring all the four stocks were equal in strength. During the honey flow the bees would fly right over the copse to save time in their haste to gather nectar. The copse was only about 30 feet away from the house.

When the bees returned home with their heavy load of nectar they would not fly over the trees but around the south side of the copse and up the narrow entrance between the house and the wattle fence. Being tired with their load of nectar and in their haste many bees would not trouble to go right up to their correct home if they came from one of the two further hives but would

enter one of the first two hives. These hives wanted extra supers to take the extra foraging field bees. The other two hives were much reduced in number but were able to carry on slowly as some of their own foragers would return faithfully to their own home. The overcrowded hives had chiefly adult field bees intent on storing honey. Neither of the four stocks required attention except the adding of supers when required. The hives were not opened for the whole of the summer and there was no fear of swarming.

EXTRA HIGH FLANGE TO BEE PORTER BOARDS.

The Bee-porter board used in a house apiary must have extra high flange or frame at the top and bottom for removing supers to avoid killing or drowning bees in the honey of the broken brace comb, and to prevent choking the bee exit through the tin bee porter escape. In a house apiary owing to the warmth and even temperature the bees are able to produce extra honey comb and usually have plenty of brace comb above the top bars

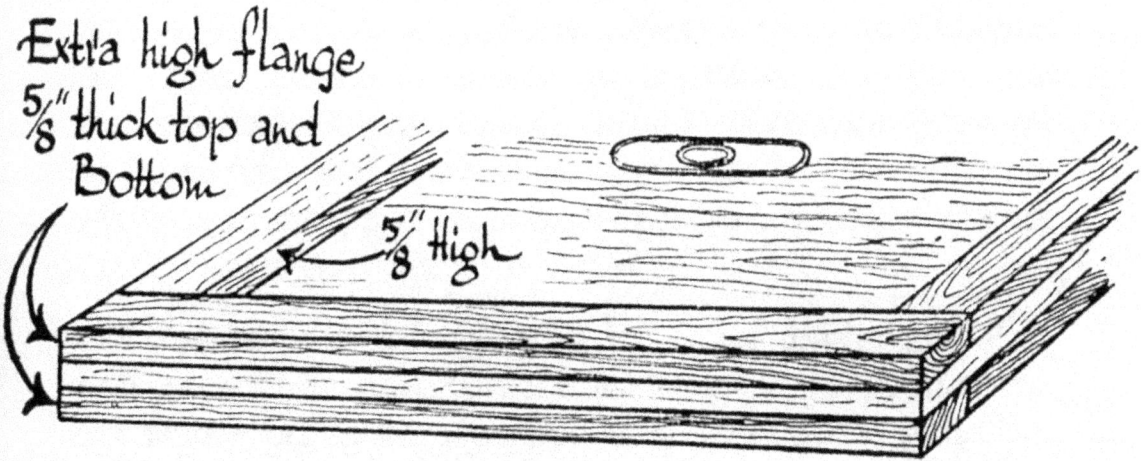

Extra high flange
5/8" thick top and
Bottom

5/8" High

and on the bottom of frames. Some may consider this a detriment, but it is a pleasure to the beekeeper to see the brace comb filled with honey and to know the bees are contented with the extra warmth, so different from the bees in outdoor hives feeling the cold when the weather changes and sometimes having to leave their supers for the night. The wet brace comb honey is not lost because the bees soon lick it up. These facts should convince the reader that bees can produce better finished shallow frames and sections of honey in a house apiary.

COMB CAPPINGS.

Another important advantage of the house apiary is that one can feed back to the bees all the wet wax cappings left over from the honey extracting at the end of the season, without any fear of attracting robber bees. When giving the bees 14 lbs. of wet cappings the bees lick up 7 lbs. of honey and leave 7 lbs. of dry beeswax in small flaky bits. Only $1\frac{1}{2}$ or 2 lbs. of wet cappings are given to one hive each day and the dry wax should be removed before adding another supply of wet cappings otherwise some drops of honey may fall into dry wax tray. In the general way beekeepers lose all the honey when washing the wet wax cappings, so here is another good dividend from the extra cost of the house.

When giving back the wet cappings to the outdoor hives in the open, great care is necessary to avoid serious robbing, especially if the beekeeper does not know the correct way to use the feeder. For this reason many beekeepers fear to use the wax cappings feeder. I am one of these, I bought one 16 years ago with printed instructions how to use it. Even some of my beekeeper friends could not understand how to use it. I only tried it once, so it has not been used for 14 years. Since then I made one of simpler design suitable for use in the house apiary as shown here in the sketch and it has been a great success.

There is no fear of attracting robbers while adding the wet cappings and removing the dry wax, the box, as can be seen in the drawing has a loose glass cover easy to remove, to add the cappings each day. With the glass cover the bees can be seen inside licking up the honey from the wet wax. This is another pleasure to the bee-keeper and his visitors in the house apiary to see the bees through the top glass cover at work.

The drawing gives details to show how the wax capping box stands immediately on top of the thin covering or enamelled American cloth over the brood frames with one corner turned back to be exactly under the hole in the bottom of wax box to allow the bees to go to and fro from the brood chamber and the wet wax cappings. The wax box should be the suitable size to stand on the frame work of a bee porter board so that it is possible to place the wax box on this at once after removing the supers and with the tin shutter opened to allow the bees to go through. One Autumn after feeding the bees with the wax cappings and with the wax box on the bee-porter board this was all left on for two weeks after the feeding was finished and the bees covered up all cold metal as shown in photograph of a tin bee-porter, evidently expecting it to be left on for the winter. This shows that they do not like touching cold metal in their home.

FEEDER BOX FOR HOUSE APIARY

BROWN PAPER EDGES TO
PREVENT CUTTING FINGERS

WOOD LEDGE FOR
RESTING WIRE
NETTING FOR
HOLDING WET
CAPPINGS

WET CAPPINGS
ON ½" NETTING

SHUTTER FOR
REMOVING TRAY
OF DRY WAX

BODY OF HIVE

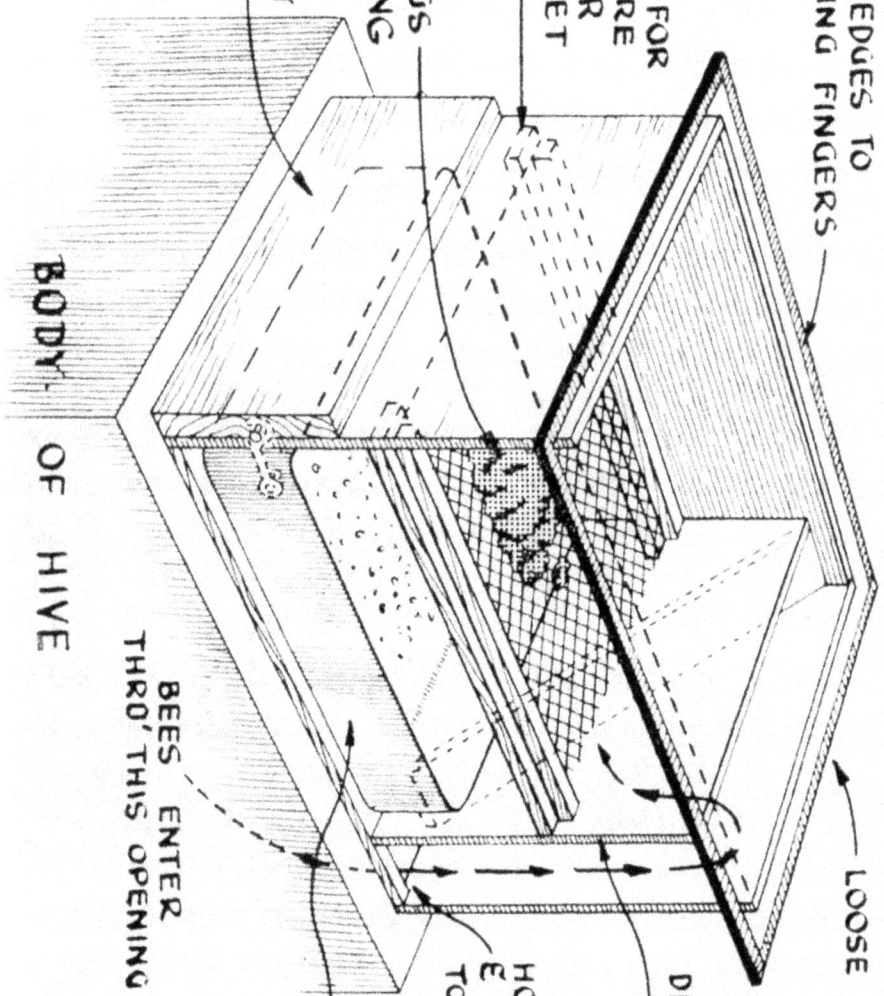

LOOSE GLASS COVER

DIVISION
WALL

HOLE ENTRANCE
& EXIT (3" x 2")
TO BODY OF HIVE

TIN TRAY FOR
DRY WAX PIECES

BEES ENTER
THRO' THIS OPENING

STARVING BEES IN WINTER.

An outdoor stock of bees with plenty of stores in the hive has been known to die in hundreds through starvation although with sufficient honey but owing to the cold weather the bees were unable to move to feed themselves. By placing the hive in a house apiary the bees are soon warm enough to look after themselves and also to remove the dead to the hive entrance. All these dead starved bees would have been saved if the stock had been kept in a house.

BEE-PROOF CUPBOARD.

It is very convenient to have a bee-proof cupboard in the house, especially in an out-apiary, to store one or two supers ready to place on hives, or to store supers of honey taken off before there is means of carrying them home for extracting. Also any frames with a little honey taken out of a hive for various reasons such as, when uniting two hives, one or two frames that may be over can be placed in the bee-proof cupboard safely out of the way of wasps.

PRECAUTION AGAINST THE CLEVERNESS OF WASPS.

It may be interesting to say that wasps have stronger powers of scent than bees. Although wasps will not fly in the darkness of the house apiary, should the beekeeper during the month of August leave a spot of honey comb or syrup exposed in the dark house the wasps are able to scent it from outside the house and will crawl in under the door or through any small place possible and walk along the floor or walls to find the honey or syrup in the dark. The bee has not such powers of scent as the wasp. Even if the bee had she would not be so bold as the wasp to go into the dark house. This is why the house apiary is such a safeguard against robber bees. It is strange that bees are so afraid of the semi-darkness of the house and yet their home inside the hive is very dark.

MATING FLIGHT OF QUEEN BEES AND WASPS.

Millions of mating flights take place out in the open every summer and autumn. Beekeepers very seldom witness these matings. I have been very fortunate to witness four of these. The first one was owing to one of my house apiaries having a seven foot wall about seven feet away from the front of the house. When the virgin queen left her hive for the mating flight she had difficulty in flying away through the crowd of drones in front of the house and those high up above the wall. She was forced to go to the left and was caught quickly before she had risen six or seven feet. I was outside the house, at the end the way she happened to fly with the drones following her and I saw the mating flight easily. There was a high laurel hedge in the background. I did not know at this time (twenty years ago) about the great controversy amongst experts on their theory of how the queen bees mated. In the above case the queen was caught by one of the drones by her hind feet trailing behind her while flying. She then hovered in a stationary position in mid-air similar to the hovering of a hawk, but only for a moment and with a drone hanging on to her feet with his own. He was upside down while the queen was hovering, she taking the whole weight of the upside down drone. The queen then curved her abdomen downward and the drone raised his upwards. The two

The Domestic Honey Bee's mating flight.

abdomens met between their feet and completely united in a quick second. They are now in the " S " shape. They both now started flying and being upside down to each other causes them to do a catherine wheel flight very quickly through the air. This part of the flight is very difficult to see owing to the swift spin. It is doubtful if anyone would see this if one had not first seen the beginning of the mating. In the case of the wasp the catherine flight is much slower and more of a clumsy wobble roll-over owing to the queen wasp being three times larger than the male. The roll-over looks like a ball of thistle-down seeds about two thirds the size of a tennis ball or less than a golf ball floating in the air which eventually falls to the ground and is lost in the grass or foliage, except if one happens to be at the spot, one will then see the queen wasp lying on her back with the small male on top and the two in the shape of the letter " S " for a few moments.

Wasps' mating time is about the end of September so this is the time to watch for the thistle-down seed floating in the air. Perhaps it will turn out to be the mating flight of the wasp.

The end of the wasp's mating flight.

The mating of the domestic bee is different from that of the wasp, or anything else. It is considered unique in the insect kingdom. At the end of the catherine wheel flight of the domestic bee, one can hear a slight explosive noise. It is assumed that this explosion is the drone's abdominal air vessels bursting in the abdomen rupturing the drone. After this the drone falls to the ground to die but the queen frees herself with her wings and returns home to her hive with a portion of the male organ showing and hanging from her abdomen. With the wasp it is different as there is no noise of an explosion and the male wasp is able to free himself and fly away and is not ruptured.

I am afraid our British Bee Experts are wrong when they condemn the house apiary. It is because of their condemnation that this booklet has been written. The house apiary is a great blessing both for the bees and for the comfort of the beekeeper, and it is hoped that enough has been said in this booklet to benefit thousands of beekeepers.